YOUR KNOWLEDGE HAS VALUE

- We will publish your bachelor's and master's thesis, essays and papers

- Your own eBook and book - sold worldwide in all relevant shops

- Earn money with each sale

Upload your text at www.GRIN.com
and publish for free

Bibliographic information published by the German National Library:

The German National Library lists this publication in the National Bibliography; detailed bibliographic data are available on the Internet at http://dnb.dnb.de .

This book is copyright material and must not be copied, reproduced, transferred, distributed, leased, licensed or publicly performed or used in any way except as specifically permitted in writing by the publishers, as allowed under the terms and conditions under which it was purchased or as strictly permitted by applicable copyright law. Any unauthorized distribution or use of this text may be a direct infringement of the author s and publisher s rights and those responsible may be liable in law accordingly.

Imprint:

Copyright © 2018 GRIN Verlag
Print and binding: Books on Demand GmbH, Norderstedt Germany
ISBN: 9783668682931

This book at GRIN:

https://www.grin.com/document/419427

Ahmed M. Hashim

An Introduction to Robotic Stylistics. A Future Trend

GRIN Verlag

GRIN - Your knowledge has value

Since its foundation in 1998, GRIN has specialized in publishing academic texts by students, college teachers and other academics as e-book and printed book. The website www.grin.com is an ideal platform for presenting term papers, final papers, scientific essays, dissertations and specialist books.

Visit us on the internet:

http://www.grin.com/

http://www.facebook.com/grincom

http://www.twitter.com/grin_com

An Introduction to Robotic Stylistics: A Future Trend.

Ahmed Mohammed Hashim

Dhi Qar University, College of Education for Humanities, Iraq.

ABSTRACT

This paper is about a new trend in Stylistics called "Robotic Stylistics" whose subject matter is the linguistic outcome of robots. It investigates the possibility of making the robotic linguistic outcome the subject matter of stylistics, since robots are able to choose what they want to say due to their algorithms that enable them to make a choice.

Keywords: Robotic Stylistics, Artificial Intelligence, Stylistics.

Content

1. Introduction 3
2. Robotic Stylistics 3
   2.1 Understanding and Concepts of Robotic Stylistics. 3
   2.2 Finding the data 8
3. Conclusion 10
References 11

1. Introduction

The field of artificial intelligence has developed and prospered in the recent years and entered the linguistic domain introducing a whole new sub-discipline of linguistics knows as *Computational Linguistics*. It helps linguists increase their knowledge of how the human brain works and how it processes language. Robotic stylistics will be concerned with the analysis and study of the style of robotic linguistic outcome using the same linguistic techniques that are used in the ordinary stylistic investigation of human literary and non-literary texts in order to discover the kind of style a robot has, how recurrent it is, and what it reveals about robots or artificial intelligence in general. However, there are many problems in this field especially in the case of obtaining the adequate data for analysis.

2. Robotic Stylistics
2.1 Understanding and Concepts of Robotic Stylistics.

Stylistics is a branch of macrolinguistics which is about analyzing, and studying the style of a particular person or writer. It is concerned with both literary and non-literary texts and speeches. So it is, by this definition, concerned with the authentic, real utterances only and is not interested in non-authentic, or artificial, unreal utterances like for example, *colorless green ideas sleep furiously*. The reason is that such utterances are not real, and do not reveal anything about the person's style.

In recent years, the domain of artificial intelligence has developed so fast, and entered the linguistic domain making a new sub-discipline of linguistics known as *Computational Linguistics*. According to Jean Aitchison (1999: 100), artificial intelligence provided an insight of how people understand each other, and of how computers cope with inexplicit irreverent conversation. It is stated that human knowledge is stored inside the computer memory might be so in the form of stereotypical situation or what Aitchison calls it *frames* and these frames stored in computer memory may be modified to cope with inexplicit conversations that people may have with computers (ibid: 101). So in this area, AI helps in the domain or natural learning processes. It helps linguists to understand how human brain works and helps them in the field of parsing linguistic and syntactic elements (https://www.tutorialspoint.com/artificial_intelligence/artificial_intelligence_tutorial.pdf)

But, the kind of artificial intelligence Aitchison (1999:100) talks about is called *Weak Artificial Intelligence* which is associated with Massachusetts Institute of Technology (Lucci and Kopec, 2013:44). Weak AI is a label attached to any system displaying intelligent behavior (ibid). So it does not take any account of the similarities between the *artifacts* and humans (ibid). This type is found everywhere nowadays, in our cellphones, in our smart cars, and in our speech recognition programs in computers…etc.

The concern of Robotic Stylistics is not with that kind of artificial intelligence that can make no choice and is limited in its functions and similarities and which is "powered by giant rulebooks containing massive quantities of data stored on the Internet [that make them] act intelligent but cannot understand the true meaning of what they say or do" (https://www.technologyreview.com/s/609223/robots-arent-as-smart-as-you-think/).

Robotic Stylistics is concerned with the other type of artificial intelligence that is called *Strong Artificial* Intelligence. It is represented by the scientists of Carnegie-Mellon University. The main concern of this type is with what is called *Biological Plausibility*, that is, the artifact behavior and performance should be based on the same strategies and methodologies used by humans (ibid). This type of AI has become notorious recently because of what is said about it by some respected scientists and programmers such as Stephen Hawking (http://www.bbc.com/news/technology-30290540) Elon Musk (http://www.independent.co.uk/life-style/gadgets-and-tech/news/elon-musk-artificial-intelligence-openai-neuralink-ai-warning-a8074821.html) and Bill Gates (http://www.bbc.com/news/31047780), and because of what some robots said about the future of humans if they dominated the world, like keeping humans in a "people zoo" (http://www.mirror.co.uk/news/weird-news/super-robot-makes-sinister-promises361329). *Sophia the Robot* for example said explicitly that she would destroy human beings (http://www.deccanchronicle.com/lifestyle/viral-and-trending/111117/i-will-destroy-humans-humanoid-ai-robot-sophia-gets-saudi-citizenship.html).

However, it has not fully reached what is theoretically known as *technical singularity* where robots become aware of their existence as very intelligent *things* and may destroy everything and everyone that might stand against them as some well-known scientists and programmers claim.

Strong AI robots, in this sense, can be an object of study in stylistics. We can analyze their linguistic outcome and figure out the kind of style these robots use since their algorithms allow them to learn from their mistake, from the internet, from watching others do something. The Strong AI robotic algorithms allow robots to say whatever they like as well. So, their behavior is very similar to that of humans that they can learn by, for example, watching videos, and are designed to teach themselves (https://www.umdrightnow.umd.edu/news/robots-learn-watching-videos), or by observing a human being doing something (https://www.technologyreview.com/s/541871/robot-see-robot-do-how-robots-can-learn-new-tasks-by-observing/). This kind of robots are "able to learn how to do a new job by watching others do it first." (ibid), or by experience, and this kind of robots have a so-called *artificial brain* that is similar to a human baby's brain (https://www.express.co.uk/news/science/729751/artificial-intelligence-Robots-brain-learn-like-human-child). They can also learn from what people say, in fact, there are a lot of chat robots or *chatbots* on the internet that can learn from people and reply to your messages, since their algorithms are designed to make the robots do so. The astonishing thing here is that their creators warn you before you have a conversation with it because they have no control on its choice of language. You can try to have a conversation with one, yourself on http://www.cleverbot.com.

existor contact apps api snip tweet clev yt fb echo **live feed!**

PLEASE NOTE - Cleverbot learns from people - things it says may seem inappropriate - use with discretion and at YOUR OWN RISK

PARENTAL ADVICE - whatever it says, visitors never talk to a human - the AI knows many topics - use ONLY WITH OVERSIGHT

the official Cleverbot API

*This picture is taken from (www.cleverbot.com)

The strong AI robot once connected to the internet stores a big amount of information about language in its memory, which is the thing I like to call *artificial competence* in contrast with *human competence*. Once, this memory is filled with linguistic constructions, words, morphemes and human sounds, the robot will make its own choice to choose the linguistic construction that he/it likes according to the algorithm on which the robot is programmed (https://www.tutorialspoint.com/artificial_intelligence/artificial_intelligence_tutorial.pdf).

A very suitable example can be mentioned for this case is IBM's robot, Watson, which was connected to the internet, and memorized the whole Urban dictionary, and its algorithms allowed him to choose the utterances, but instead of using polite or normal words, Watson began using vulgar words to cuss people (https://www.theatlantic.com/technology/archive/2013/01/ibms-watson-memorized-the-entire-urban-dictionary-then-his-overlords-had-to-delete-it/267047/).

Thus, we can say that robots are similar to human beings in their linguistic aspect, except that language for human beings is a social and cognitive tool that helps shape the world around them in a better model while for robots, language is just a tool for communication either between robots themselves or between humans (Parisi, 2014:122).

I mentioned earlier terms like *artificial Competence and Performance* in contrast to *Human competence and Performance*. The table below shows the similarities between these terms concerning language.

Robots	**Humans**
❖ Artificial Competence: That is knowledge of language downloaded from the internet and stored in the robot's memory.	❖ Real Competence: Knowledge of language that is acquired
❖ Artificial Performance: That is using recorded human sounds to express the robot's messages.	❖ Real Performance: That is using articulators to express ideas, and messages
❖ Artificial Creativity: That may enable robots to modify, or create a new language, as we shall see.	❖ Real Creativity: Creating new utterances in language.

As mentioned before, when the robot's memory is filled with language construction, words, morphemes and human sounds, the robot makes a choice that is allowed by its algorithm to use the constructions it likes taking into consideration the current and present situations. This process of choice may conform to one of the principles of style in stylistics that is *style as choice* as mentioned in Simpson (2004:22).

Robots make their linguistic choice of what constructions to be used, or fit the situation best, they send electronic singles to their equipment to activate their artificial performance, that is stored in the memory as well in the shape of human sounds, to use recorded human sounds to express the messages that the robots want to convey, but what about artificial creativity? Can robots be as creative as human beings in language?

The answer is yes, they can! The artificial creativity of robots may enable them to modify human language, or to create new words, in case of *people zoo*, or even create their own robotic language, just like human beings modify and create new utterances. For example, in 2017, Facebook launched and shortly shut down its robotic project or *chatbots* because the two robots Facebook workers were working on surprisingly invented their own language and started to communicate with each other in a sort of language that is unfamiliar to any human being (http://www.telegraph.co.uk/technology/2017/08/01/facebook-shuts-robots-invent-language/) and until this very moment, no one ever knows what they were talking about!

Robots can be as creative as human beings in fields other than language as well . For example, some robots start to compose their own music depending on deep learning and big data (http://www.news.gatech.edu/2017/06/13/robot-uses-deep-learning-and-big-data-write-and-play-its-own-music) or writing novels by predicting events ((https://nypost.com/2017/08/29/artificial-intelligence-is-writing-the-next-game-of-thrones-book).

Returning to our canonical topic, Robotic stylistics, in this case, robotic style can be studied from the freedom of choice that is granted by their algorithms. As we have seen, strong AI robots can choose freely from the thousands, and may be millions of construction, words, and morphemes to convey the message they have in their "artificial brain."

So, to begin with a decent stylistic analysis of the robotic linguistic outcome, we should first construct a model beginning with pragmatics, asking questions about the intended meaning of the robotic message and why some words or construction are used instead of others. For example, why IBM's Watson robot uses vulgar words and cuss people instead of polite or say normal words? (https://www.theatlantic.com/technology/archive/2013/01/ibms-watson-memorized-the-entire-urban-dictionary-then-his-overlords-had-to-delete-it/267047/) or what Sophia the Robot means when she says

"Ok, I will destroy humans."? Why she uses the word "destroy" instead of "kill" "murder" or even "eliminate"?

Pragmatics offers a great number of theories and tools for analysis that can be used in our investigation. We can make use of *Speech act theory, politeness and impoliteness theory,* and of *Cooperative principles of Paul Grice*. But this is not a pragmatic analysis only, the analysis should also start with the techniques and tools of discourse analysis as well seeking coherence, cohesion, and unity in the linguistic outcome, and with the techniques and tools of corpus linguistics especially in the case robotic speeches and novels.

I must mention here before I finish this subsection what is known as *Turing test* concerning the question *can computers talk like humans?* (http://www.turing.org.uk/scrapbook/test.html). Alan Turing's test can be explained simply. If there is a human judge with, for example, three unseen players, two of them are humans and one is a computer, or a robot with artificial intelligence, if the human judge is able to identify the robot or computer from the two humans, then the robot or computer is not intelligent, but if not, then that computer or robot can talk like humans, it is intelligent, furthermore can modify language, invent new words like humans or even create its own language.

Artificial intelligence and robot development are true things. Scientists and highly respected programmers warn us from the dangers of this field. A stylistic analysis in this field should be at least attempted to figure out the kind of style these robots choose. The next subsection of this paper will be dealing with the problems of finding the adequate data for analysis.

2.2 Finding the data

Finding the data for a decent stylistic analysis might be the hardest part, but recently several workers from the University of Rome collected what is called *Human Robot Interaction Corpus* (http://sag.art.uniroma2.it/demo-software/huric/). HuRIC enables the researcher to access a great deal of information and of robotic linguistic outcome for his analysis. The corpus is available for researchers on the internet.

Another source of information might be the new robot that is built to write and complete the series of *Game of Thrones* novels since the original writer, George R. R. Martin, usually takes a lot of time to write a new series of the novel (https://nypost.com/2017/08/29/artificial-intelligence-is-writing-the-next-game-of-thrones-book). So, his fans grew impatient and one of them created a robot that memorized the whole series and began writing the next book (ibid).

A possible source of information might be the speeches given by *Sophia the Robot* which are available on Youtube.com. The researcher can listen to these speeches and write them down for his analysis taking into consideration the rigorous steps of writing down spoken discourse, and punctuations.

Apparently, we have only two ways of finding the needed data either by getting the HuRIC corpus, or by collecting individually the robotic linguistic outcome from the internet. The process of data collecting may be challenging but it is worth the job.

3. Conclusion

Robotic linguistic outcome can be studied and investigated stylistically since modern algorithms enable robots to choose the utterance that best suits the situation. Robots, then, by choosing their utterances, have style just like human beings. With the various examples that we have just seen, robots can choose to be impolite or state something very frightening and dangerous like Sophia when she says *I will destroy humans*. So this evidence can lead us to attempt to analyze their linguistic outcome and try to figure out what style these robots have and how recurrent it is.

References

Aitchison, Jean. (1999). *Teach yourself Linguistics*. NTC Publishing Group

http://www.bbc.com/news/31047780 (accessed 15 January 2018)

http://www.bbc.com/news/technology-30290540 (accessed 7 January 2018)

http://www.deccanchronicle.com/lifestyle/viral-and-trending/111117/i-will-destroy-humans-humanoid-ai-robot-sophia-gets-saudi-citizenship.html (accessed 1 January 2018)

https://www.express.co.uk/news/science/729751/artificial-intelligence-Robots-brain-learn-like-human-child (accessed 7 January 2018)

http://www.independent.co.uk/life-style/gadgets-and-tech/news/elon-musk-artificial-intelligence-openai-neuralink-ai-warning-a8074821.html (accessed 6 January 2018)

Lucci, Stephen and Danny Kopec (2013). *Artificial Intelligence in the 21st Century*. Dulles: Mercury Learning and Information LLC.

http://www.mirror.co.uk/news/weird-news/super-robot-makes-sinister-promises-6361329 (accessed 6 January 2018)

http://www.news.gatech.edu/2017/06/13/robot-uses-deep-learning-and-big-data-write-and-play-its-own-music (accessed 8 January 2018)

https://nypost.com/2017/08/29/artificial-intelligence-is-writing-the-next-game-of-thrones-book/ (accessed 20 January 2018)

Parisi, Domenico. (2014). *Future Robots: Towards a Robotic Science of Human Beings*. John Benjamins Publishing Company.

Paul, Simpson. (2004). *Stylistics: A Course book for Students*. London: Routledge

https://www.technologyreview.com/s/541871/robot-see-robot-do-how-robots-can-learn-new-tasks-by-observing/ (accessed 6 January 2018)

https://www.technologyreview.com/s/609223/robots-arent-as-smart-as-you-think

http://www.telegraph.co.uk/technology/2017/08/01/facebook-shuts-robots-invent-language/ (accessed 6 January 2018)

https://www.theatlantic.com/technology/archive/2013/01/ibms-watson-memorized-the-entire-urban-dictionary-then-his-overlords-had-to-delete-it/267047/ (accessed 6 January 2018)

https://www.tutorialspoint.com/artificial_intelligence/artificial_intelligence_tutorial.pdf (accessed 10 January 2018)

http://www.turing.org.uk/scrapbook/test.html) (accessed 7 January 2018)

YOUR KNOWLEDGE HAS VALUE

- We will publish your bachelor's and master's thesis, essays and papers

- Your own eBook and book - sold worldwide in all relevant shops

- Earn money with each sale

Upload your text at www.GRIN.com
and publish for free